The
Housefly

The Housefly

by Heiderose and Andreas Fischer-Nagel

A Carolrhoda Nature Watch Book

Carolrhoda Books, Inc./Minneapolis

*Many thanks to Professor Ralph W. Holzenthal,
Department of Entomology, University of Minnesota,
for his assistance with this book*

To our children, Tamarica and Cosmea Désirée

This edition first published 1990 by Carolrhoda Books, Inc.
Original edition copyright © 1988 by Kinderbuchverlag KBV Luzern
AG, Lucerne, Switzerland, under the title DIE STUBENFLIEGE.
Adapted by Carolrhoda Books, Inc.

LIBRARY OF CONGRESS CATALOGING-IN-PUBLICATION DATA

Fischer-Nagel, Heiderose.
[Stubenfliege. English]
The Housefly / by Heiderose and Andreas Fischer-Nagel.
p. cm.
Translation of: Die Stubenfliege.
"A Carolrhoda Nature Watch Book."
Includes index.
Summary: Describes, in text and illustrations, the physical characteristics, habits, natural environment, and relationship with humans of the housefly.
ISBN 0-87614-374-5 (lib. bdg.)
1. Housefly—Juvenile literature. (1. Housefly. 2. Flies.) I. Fischer-Nagel, Andreas. II. Title.
QL537.M8F5713 1990
595.77'4—dc20 89-32365
CIP
AC

Manufactured in the United States of America

1 2 3 4 5 6 7 8 9 10 00 99 98 97 96 95 94 93 92 91 90

It's hot, and you hear the familiar buzz of a fly as it swarms around your head. Your first reaction may be to swat it away. But wait! Flies can be fascinating, and it may be a good time to take a closer look.

In this book, you'll have a chance to study a very familiar fly—the housefly. Although you've seen this flying acrobat for years, you'll now be able to examine how much it has in common with other insects, as well as how unusual and interesting it is.

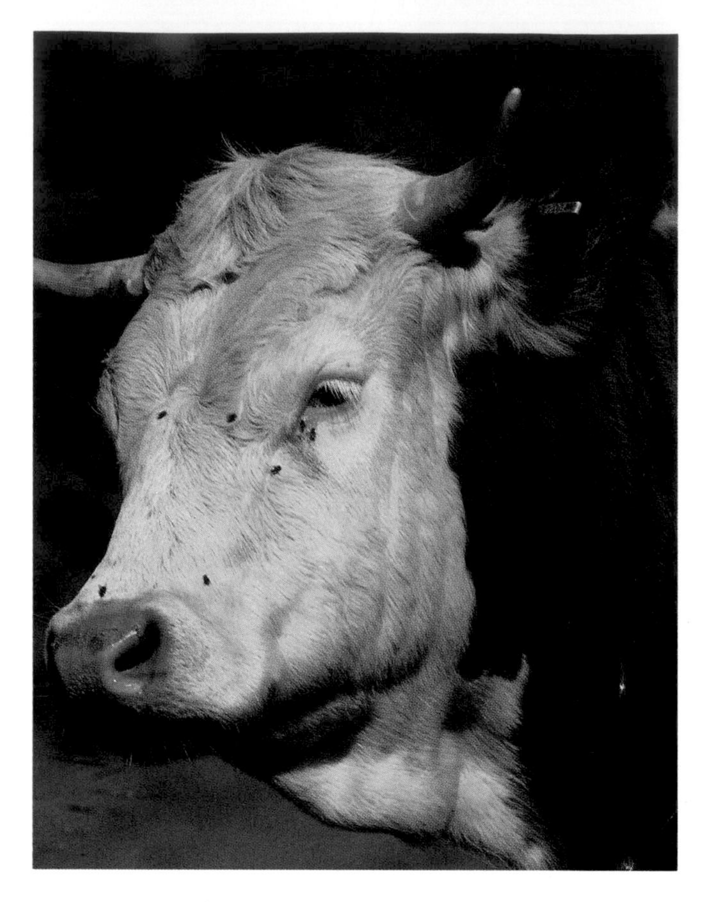

Most people think of flies as pests, but if you had lived in ancient Eygpt, you may have felt differently. In 3500 B.C., the fly was a symbol of bravery. Flies made of gold were presented to soldiers who had been courageous in battle.

In the past, flies were also looked upon as signs of good fortune. The more food and livestock people had, the more flies surrounded them.

The housefly is a member of the scientific group called true flies, or *Diptera*. In Greek, *di* means two, and *ptera* means wings. Although many insects have "fly" as a part of their names, not all of them are members of *Diptera.*

True flies have only two wings—one on each side of their bodies. This gives them greater flying abilities than most four-winged insects because they have fewer wings to balance and control. That's why some *Diptera* can do amazing tricks, such as flying in zigzags or even backward through the air.

Originally, the ancestors of true flies had a second pair of fully developed wings. Over the course of millions of years, however, these wings shrank in size. Now they are just two small spoon-shaped rods, which are called **halteres.** They are found behind the wings. One is circled in the picture below.

Although halteres are quite small, they are very important. These rods shift up and down during flight, balancing the *Diptera* as they twist and turn in the air. With halteres and wings working properly, somersaults and other astonishing feats are no problem for these agile flyers.

There are so many **species,** or types, of *Diptera* in the world that scientists have trouble keeping track of all of them. It is estimated that there are presently between 85,000 and 100,000 species in the world. Approximately 17,000 of these species live in North America alone.

Diptera species have been categorized into 108 families. The housefly, or *Musca domestica,* is in the family *Muscidae* along with stable flies and other relatives.

Houseflies can be found almost anywhere there are people. And as with people, the colder it gets, the fewer houseflies you find. So if you are on the lookout for houseflies, you'll have the best luck in the summer, watching them swarm around food, beverages, animals, garbage, and manure.

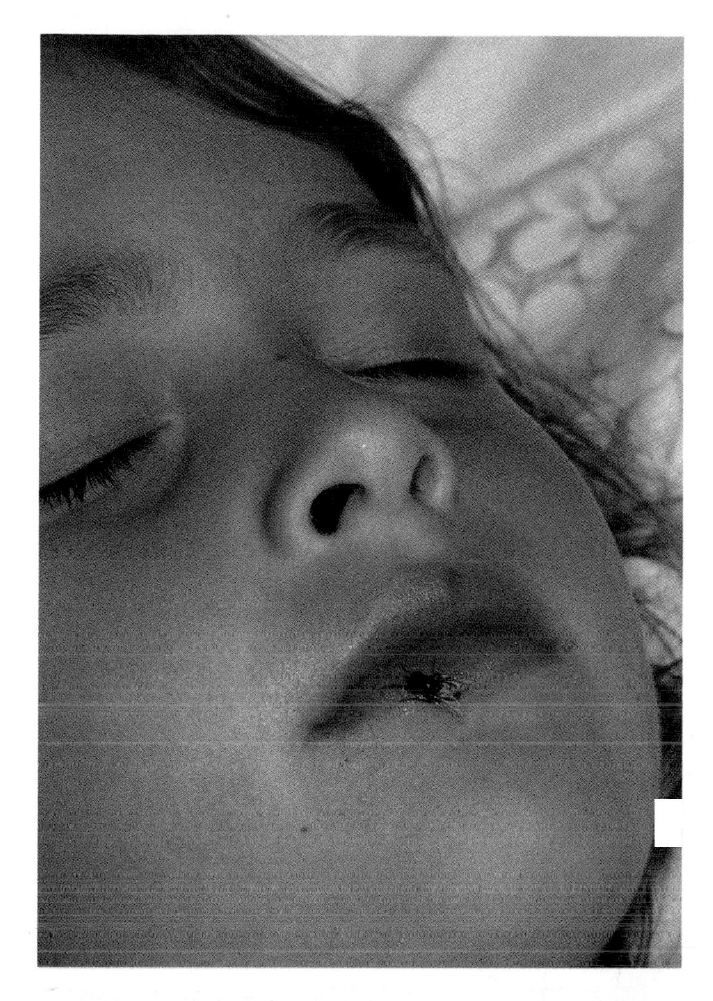

As you can see in the life-size picture below, the housefly is only 3/10 of an inch (7.62mm) long, and it weighs even less than you may expect. It would take nearly 2,000 flies weighed together to equal 1 ounce (70 flies for 1 gm).

If you examine a housefly under a magnifying glass or microscope, you will see that, like all other insects, it has three sections to its body: the head, the thorax, and the abdomen.

The head of a housefly has two large, red-brown eyes. Beneath them are two **antennae,** or feelers, and the mouth parts.

The **thorax** is the middle section. Extending from the thorax's underside are six hairy legs, and when the fly is at rest, two soft wings lie over its back.

The end section is the **abdomen.** The abdomen carries the majority of the fly's digestive and reproductive organs.

A housefly is light because it is not weighed down by bones. Like all insects, the housefly has an **exoskeleton** (EK-so-SKEL-uh-ton), or outer body casing, that protects and supports its muscles and soft insides.

One of the substances that makes up an exoskeleton is **chitin** (KAI-tin), which cannot stretch much after it has hardened. When an insect outgrows its exoskeleton, it **molts,** or sheds this outer layer. The exoskeleton splits down the middle, and a soft insect crawls out. Within a few hours, the outer layer of the insect's body hardens into a new exoskeleton.

The smallest section of the housefly's body is the head, and from far away, the head seems to consist almost entirely of its two eyes. These eyes are large because they are **compound eyes**. Each compound eye is made up of between 3,000 and 6,000 simple eyes that are interconnected.

A COMPOUND EYE

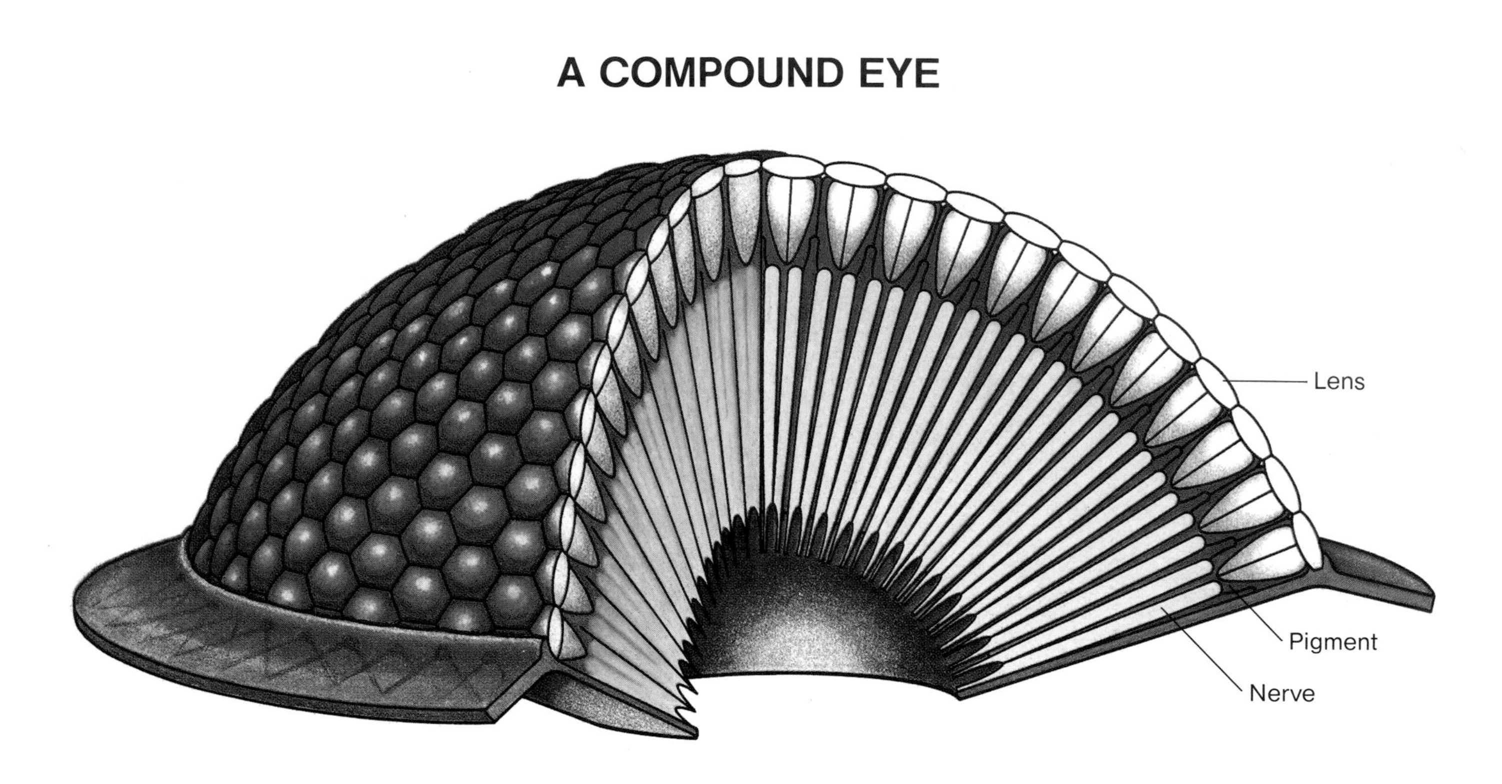

As shown in the diagram of a compound eye, the simple eyes are connected together like a honeycomb. Every simple eye has its own nerve, red-brown coloring, and six-sided lens. So a housefly sees not one smooth image but a mosaic of thousands of small pieces of a picture, all at one time.

The compound eyes may seem too large for a fly's small body, but their size is necessary for the insect's survival. An insect with compound eyes can see in many directions at once. Each six-sided lens points in a slightly different direction, so the insect has a wide range of vision. A housefly is able to see above, below, and to the sides as clearly as it can see to the front. That is why it is so difficult to sneak up on a housefly—the fly can almost see behind itself.

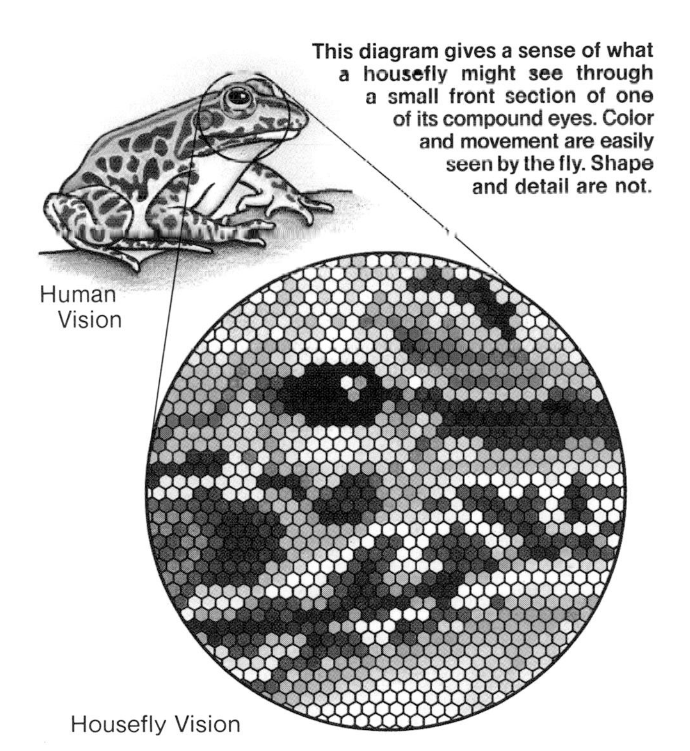

This diagram gives a sense of what a housefly might see through a small front section of one of its compound eyes. Color and movement are easily seen by the fly. Shape and detail are not.

You may wonder how flies can focus on everything at once. The answer is that they can't. Compound eyes cannot be focused. Every small image in a compound eye's mosaic is blurry.

Because houseflies never need to focus their eyes, they can see things more quickly than we can. Flies are able to detect even the slightest movements almost immediately, giving them time to escape approaching enemies.

The ability to detect things quickly also helps houseflies identify other flying insects. The wing speeds of different insects vary, and each wing beat makes its own flickering light pattern. Male houseflies often find mates by following the light patterns made by flying females.

In addition to their compound eyes, houseflies have three tiny simple eyes, called **ocelli** (oh-SEL-eye), arranged in a triangle between their two compound eyes. Scientists believe that these simple eyes help flies maintain direction in flight. As day animals, houseflies determine direction by the location of the sun, which is why they are often found on windowsills. Scientists believe the ocelli may serve as a kind of light meter, measuring the amount and intensity of light rays.

Houseflies, like many other insects, are also able to see colors. Brightly colored objects, such as flowers, can attract flies because of their color. Houseflies may even be able to detect colors humans can't see.

Colors have proved useful to scientists in their studies of houseflies. By training houseflies to lower their mouthparts for food when a certain colored light is shone on them, scientists have discovered that houseflies have some memory and learning capabilities.

Houseflies have other keen senses, but they are not all centered in their head. Houseflies can smell with their antennae, feel with their hair, listen with their legs, and taste with their feet.

The fine hairs on a housefly's antennae are highly sensitive to the chemicals in the air that make up odors. These chemicals make the hairs on the antennae twitch, sending signals of smells to its brain. The fly then acts by instinct, reacting according to whether the scent is from a possible mate, an enemy, or a food source.

Covering a housefly's legs and thorax are thicker bristles that respond to air currents and humidity, giving the housefly the ability to "touch" the air. In addition, some of the legs' bristles are sensitive to the movements of sound waves. The sound waves vibrate the bristles at different speeds and rhythms. These vibrating bristles provide the fly with a type of hearing.

You may have watched a fly land on a piece of food and then step carefully around on top of it. The fly is testing the material to see if it would be good for eating or for egg laying. The hairs on its **tarsi,** or feet, react to the chemicals in the material in a way that is similar to the way our tastebuds react to flavors in food. Thus, the fly is "tasting" with its feet. At the same time, the fly is also checking the texture and moisture content of the material.

If the material is suitable for eating, the fly will taste it again with its **palpi,** or small, antennaelike feelers on its mouth. The fly will then lower a hollow tube, called a **proboscis** (pruh-BAH-sus), out of its mouth. The proboscis is like a funnel, with two absorbent pads, called **labella,** at the end. The fly sucks the liquid up through the labella. Fine hairs around the lower edges of the labella allow the fly to taste as it eats.

Scientists call houseflies spongers because everything they eat must be a liquid that can be sucked up. Houseflies have no teeth for chewing or biting. To eat something solid, such as a sugar crystal, a housefly must suck on it. Saliva flows down the proboscis, and the digestive juices in the saliva break the food down so it can dissolve and be sucked up.

But this process can't turn everything into a liquid. Some tiny particles often remain. Thus, instead of going directly to the fly's stomach, these particles go through a separate tube to an inner sac, called the **crop**, for storage. Later, a drop of saliva with the partially digested food will flow up from the crop to the labella for a new application of saliva. This is what is happening in the bottom photo. Food particles will make this journey up and down until they are liquefied enough to go into the fly's stomach. This process is similiar to a cow chewing its cud.

Unfortunately, if a fly goes through this regurgitating process while standing on something people eat, disease can be spread. That is why it is important to keep houseflies away from food.

To people, houseflies do not seem selective in their tastes. They move back and forth between animals, food and drinks, garbage, manure, open wounds, and even the decaying bodies of dead animals. As they land on one thing and go on to the next, houseflies pick up disease-causing germs and carry them in their crops, as well as between the hairs on their feet, legs, and mouthparts.

Despite this, houseflies are extremely clean animals and spend a lot of time grooming themselves. Using their leg bristles like combs, they clean their front legs, middle legs, hind legs, heads, proboscises, abdomens, and wings.

Keeping its wings clean is essential for the housefly's survival. Each of the fly's transparent wings is made of two microthin layers of chitin that are supported by tiny veins. Even the smallest particle on one of these wings is enough to throw off the housefly's balance and flight speed. This could be a matter of life or death for the fly because it has no other defense against predators. The housefly depends on speed and acrobatics in flight to escape its enemies.

Houseflies fly at an average speed of 4.5 mph (7.24 kph). That is similar to a person four feet (1.219 m) tall traveling at about 625 mph (1,006 kph). And houseflies can fly even faster when they are being chased.

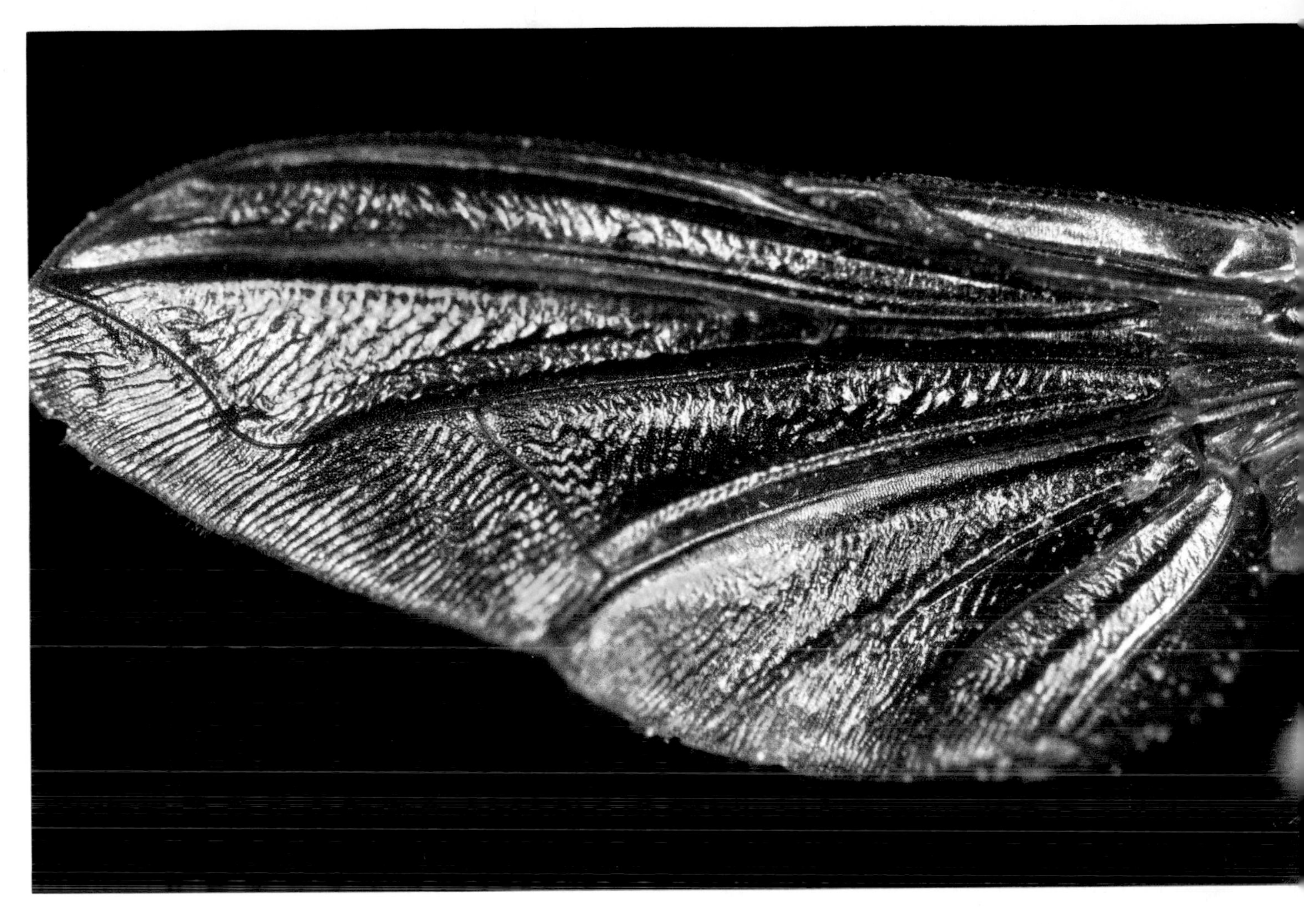

Using special sound instruments, scientists have found that the wings of a housefly beat 200 to 300 times per second. Another familiar insect, the mosquito, has wings that beat even faster—at an average speed of 600 beats per second. And the faster the wing speed, the louder the hum. That is why the hum of a mosquito is louder than that of a housefly.

Houseflies do not need a running start to fly. They beat their wings and are immediately airborne. And the wings do not stop beating until their feet touch the ground. This is an automatic reaction. If you pick up a fly by the abdomen, its wings will immediately begin to beat because its feet are not touching anything.

Have you ever wondered how a housefly manages to stick upside down on a ceiling or sideways on a cool window pane? The answer lies in the complex makeup of its leg and foot. As you can see in the right-hand photo, the fly's lower leg has three segments: the **femur**, or upper segment; the **tibia**, or lower segment; and the tarsus, or the foot, which has five segments of its own.

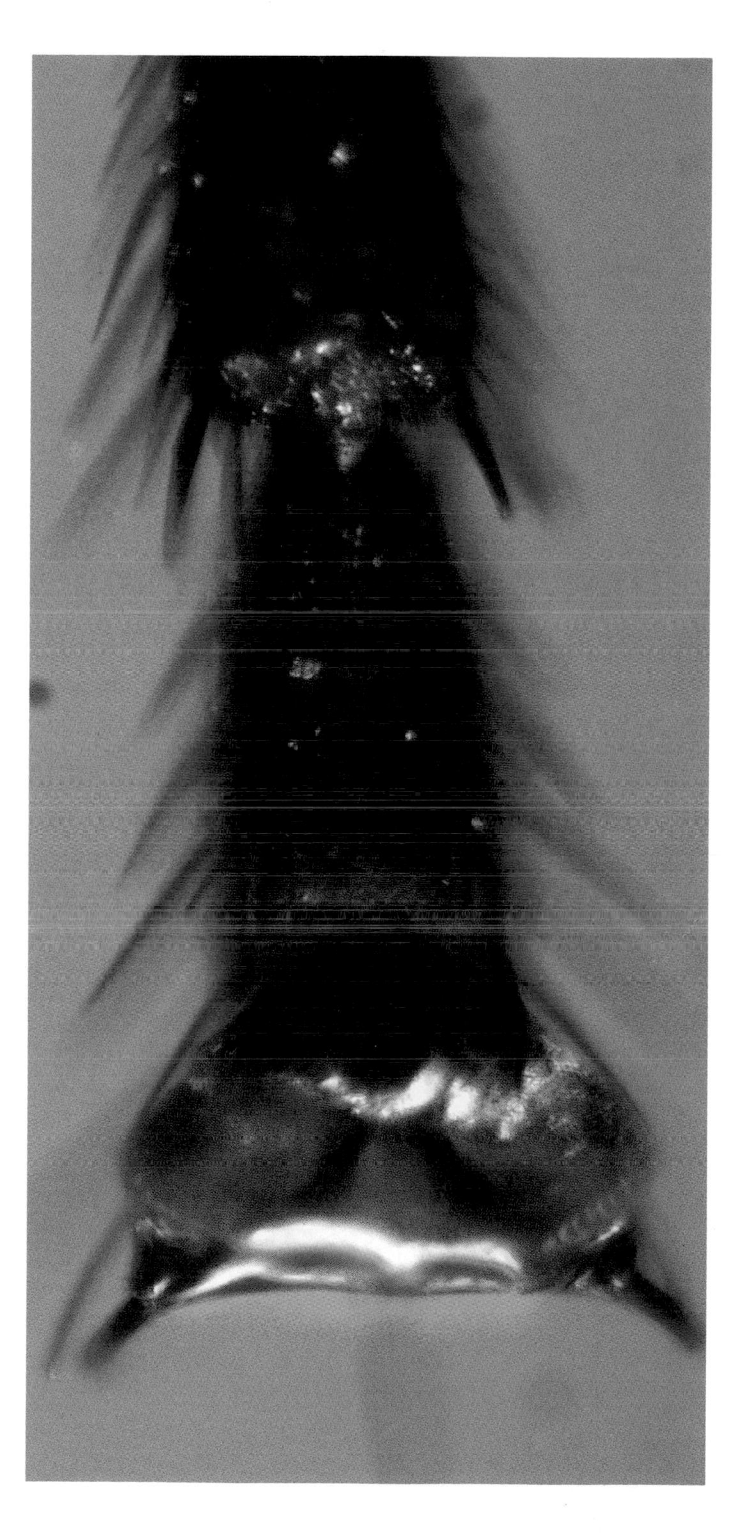

The photo to the left is an enlargement of the last segment of the fly's tarsus. Attached to it are two adhesive balls, called **pulvilli** (pull-VIL-eye), and two claws. Both are covered with fine hair. The housefly uses the claws to clasp rough surfaces and the pulvilli to stick to smooth surfaces. The pulvilli work like small suction cups, with the natural moisture of the fly's foot ensuring suction.

Whether a housefly is in the air or on the ground, it needs oxygen moving throughout its body to survive. But it does not have lungs as many other animals do. Like other insects, the housefly has small openings, or **spiracles** (SPIR-uh-kuhls), on its body. Air passes freely in and out of these holes, bringing oxygen directly to the inner organs.

Flies also have a liquid flowing through their bodies that is similar to blood, but that does not carry oxygen like human blood does. This liquid carries nourishment and wastes, and keeps the fly's inner body moist. Instead of being red and warm, the fly's blood is greenish and the same temperature as its environment.

In general structure and overall appearance, male and female houseflies are so similar that it is difficult to tell them apart from a distance. Only during the stages of reproduction do they look and act significantly different.

You may have seen two houseflies clutching each other as they fly through the air. This is the first step of the mating process. The male finds the female by her scent and wing beat. While they are both still in flight, the male often seizes the female with his legs and forces her to land. Then he climbs on her back, holding her head tightly with his front feet.

From his place on her back, the male housefly touches the female between the eyes with his proboscis. The touching of the female's head makes her stretch herself out so her **ovipositor** becomes visible at the end of her abdomen. The female's ovipositor is a tube-like organ for egg laying.

The male then lowers the back of his abdomen, and extends his **aedeagus** (ee-dee-AY-gus), or his sperm-depositing organ. He places his aedeagus into the ovipositor and deposits his sperm. This coupling can take between a few seconds and a few minutes, and it is the last contact the two flies have with each other.

The female is ready to lay her eggs four to eight days later. First she finds warm, moist, food material, such as decaying plant matter, manure, or a dead animal. She chooses her site carefully so her offspring will not have to search for food once they hatch.

Having found an appropriate material, the female walks over it, looking for cracks and tunnels in which she can hide her eggs. Using the end of her extended ovipositor as a drill, she pushes it as deep into the material as possible. She then starts pressing out long, white eggs. As the eggs pass through the end of her ovipositor, they meet the sperm still stored there. This process is called **fertilization.**

Over the course of the day, the female housefly will lay 100 to 150 fertilized eggs in different places. By scattering her eggs in various locations, the female is ensuring that one predator won't eat all of her eggs.

Larvae begin to emerge 12 to 24 hours after the eggs have been laid. Larvae are small wormlike creatures that are also called maggots. The larvae of the housefly are only slightly developed. They have small mouth hooks at their head ends, no body sections, and no legs.

The skin of the larvae is soft and slightly elastic, so it can stretch a little as the larvae grow. Larvae grow swiftly because they spend all their time eating. Many begin this stage by eating their egg casings.

The larvae have doubled in size in 1 to 1½ days after hatching, and they molt for the first time. Their skin splits across the back, and they wiggle their way out. Then the larvae begin to eat and grow some more. This molting cycle occurs two more times before the larvae reach the next stage of growth. After each molting, the larvae emerge larger and a little more developed.

Once the third molt is complete, the larvae seek a cool place to hide. They wriggle as deeply as they can into the material they are in, for their only protection against predators is to avoid being seen. Many larvae are not hidden well enough, for they are eaten during this time by birds, reptiles, and other predators.

After the larvae have settled and their bodies are at rest, their skins begin to harden and darken into tough casings. The larvae have begun the pupal stage—the intermediate stage of development between larvae and adult flies. During this stage, the **pupae** remain motionless in their pupal casings, like caterpillars in cocoons.

Inside the casings, the bodies of the pupae begin to change rapidly, developing different parts—heads, thoraxes, and abdomens; wings and legs; eyes, antennae, and bristles.

After three to six days, fully developed flies start to hatch from their casings. A kind of blister has formed on the head of each fly pupa. With this hardened point, the fly pushes its head out of the pupal casing and crawls out.

The newly hatched insect cannot yet fly. Its body is still soft and damp, and its wings are folded together. Resting from its effort, the young fly waits for its wings to dry and its exoskeleton to harden. During this time, the "breaking-out" blister shrinks back into its head.

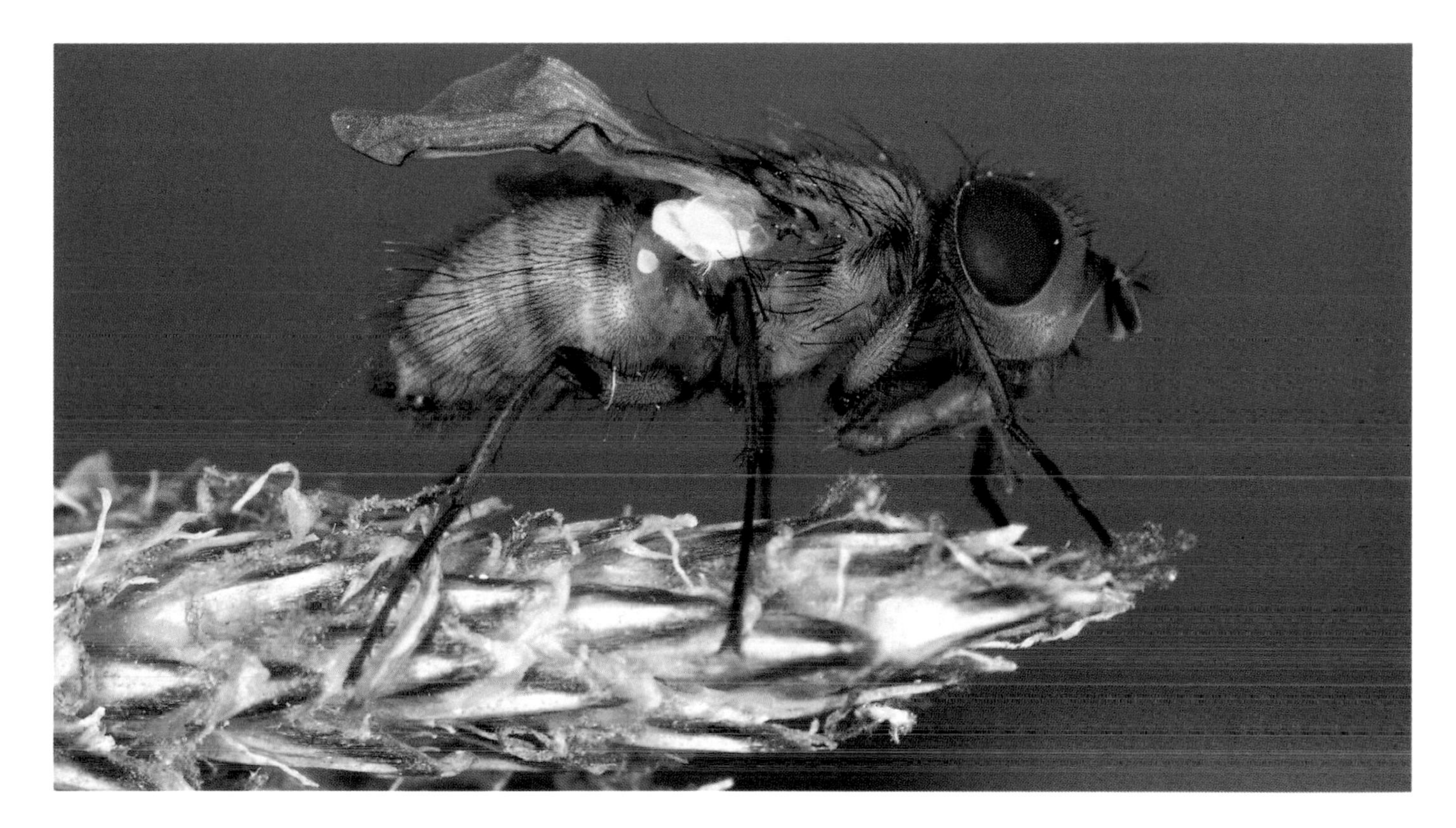

The housefly's **metamorphosis,** or change from egg to adult, can be completed in as little as 8 days, but it generally takes between 10 and 14 days. The new females are then able to mate and lay eggs within 3 to 10 days of their hatching.

The young adults will not roam far. Houseflies usually spend their lifetimes within ½ to 9 miles (.8 - 14.48 km) of where they hatched.

Houseflies have relatively short life spans. They live between two weeks and three months, with the average lifetime being 21 days. The length of the fly's life often depends on the weather—the temperature, humidity, and availability of water.

The coming of winter usually means death to the housefly. Houseflies freeze to death when the temperature falls below 32° F (0° C). At 39° F (3.9° C), the housefly is already too cold to move. When the temperature rises above 115° F (46.5° C), houseflies die from the heat.

Under certain conditions, however, a housefly may survive extreme temperatures. To protect itself, the fly's body shuts down temporarily. This bodily state is called **diapause,** and it is similar to hibernation. The fly will revive when the temperature returns to normal —if conditions haven't been too extreme. Flies may also go into diapause because of a lack of moisture or food.

Diapause comes on gradually and works to ensure that the housefly species survives from summer to summer. Flies can go into diapause at any stage of the life cycle—egg, larva, pupa, or adult. Most of the eggs laid in late summer or early fall go into diapause until spring.

Some houseflies, though, do not go into diapause during the winter. They manage to survive for a few winter weeks in an indoor environment. But these indoor flies often do not last long. They are attacked by an enemy smaller than they are—*Empusa muscae. Empusa muscae* is a microscopic fungus that infects and paralyzes flies, leaving a powdery white covering on their bodies.

Though houseflies have short life spans, they reproduce at amazing rates. Three days after a female housefly lays eggs, she is ready to mate and begin the reproductive cycle again. The average female lives long enough to lay two to six batches of 100 to 150 eggs, with approximately two weeks between each batch. Therefore, it is estimated that one female could produce over 125 billion great grandchildren if all were to survive. But this is not likely to happen. Houseflies have many enemies.

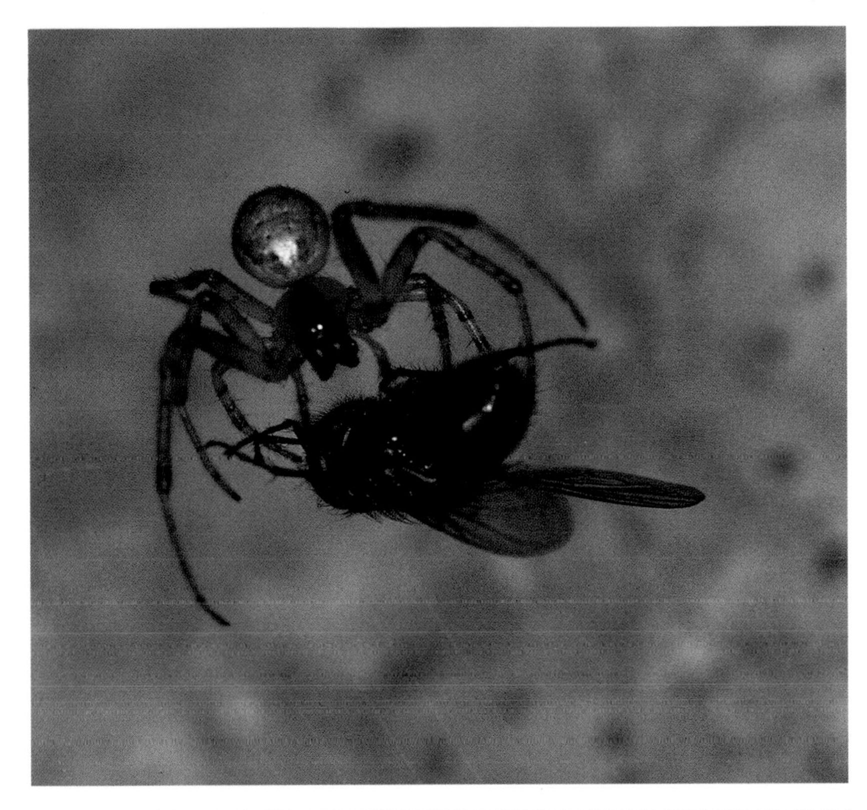

Many animals enjoy eating houseflies. But, because houseflies are so quick and agile in the air, these predators have to use special methods to catch them.

The frog, for example, unrolls its sticky tongue at a lightning speed to pluck flies out of midair.

The spider spins a fine web that cannot be seen with the fly's compound eyes. The web snags unsuspecting flies in flight.

The swallow uses quick, swooping movements to scoop up flies before they are aware of any danger.

The lizard sits without moving, waiting for flies to land on it. Then the lizard snaps them up.

People also have a variety of fly-catching methods, which they use to prevent flies from spreading disease. Houseflies have been responsible for passing on the germs for typhoid, dysentery, salmonella, cholera, diarrhea, tuberculosis, conjunctivitis, and leprosy, among other diseases.

Flypaper is a popular method used to catch flies. These long, sweet-smelling sticky strips are often hung up to attract flies. When the flies land, they are stuck in place.

Then there is the flyswatter. Flyswatters work best when they have mesh screens at the end. Like the spider's web, these screens cannot be seen with the fly's compound eyes.

Insect sprays are another option. They are convenient and efficient, but they cause problems when used improperly. Poisonous chemicals sprayed into the air often kill other insects as well as flies. And many of these insects are needed to help our gardens grow and be food for birds. The chemicals in insect sprays can also be harmful to humans and other animals.

The housefly may be the true fly seen most often, but there are other species of *Diptera* you have probably seen before.

The little housefly, or *Fannia canicularis*, is smaller than the common housefly, being about 1/5 of an inch (5mm) long. Slender and dark gray, this fly has three dark stripes across its back and pale yellow sides. Unlike the housefly, *Fannia canicularis* is a master at hovering, and when at rest, its wings extend at a straight angle from its sides.

In the family *Sarcophagidae,* the gray flesh fly is common. Often called the "buzzer" because of its exceptionally loud hum, the gray flesh fly feeds on dead flesh, helping matter to decay.

Here are two representatives of the family *Syrphidae,* or hover flies, which has about 4,600 species. As with hummingbirds, the wings of hover flies beat so quickly that we can't see them.

At one time or another, you may have been frightened by one of these flies. They look very much like bees and wasps with their yellow and black stripes. This coloring protects hover flies from predators, who mistake them for their stinging look-alikes.

And like bees and wasps, hover flies pollinate flowers as they move from bloom to bloom. The larvae of hover flies do their part in the garden as well. They save rose bushes and other plants by eating some smaller insects, such as aphids. Because of these traits, hover flies are also known as flower flies.

Blow flies, from the family *Calliphoridae,* are another common group of true flies. They lay their eggs in animals' open wounds, and their larvae feed off both infected and healthy tissue. Unfortunately, this can be both painful to the animals and damaging to the tissue.

Certain blow fly species, however, eat only the unhealthy tissue. Because of this, they have been useful in medicine. The emperor gold fly, shown below, is one of these species. Their larvae eat the infected tissue only.

In the American Civil War, doctors in the field sometimes used blow flies to deeply clean their patients' wounds. The saliva of some of these flies actually sped up the healing process.

Many flies, though, such as stinging and biting flies, seem to be known solely for the damage they cause. The tsetse fly, whose home is in Africa, often passes on sleeping sickness when it bites. This disease is nearly always fatal to humans and other animals.

The mosquito is another stinging *Diptera.* Besides making many of us miserable in the summer, it is the insect responsible for the tropical disease malaria. Unlike the germs that are accidentally carried by houseflies, the organisms that cause malaria depend on mosquitos to develop. They grow within mosquitos and are passed on to people when the insects sting them.

Although flies can do a lot of harm, they also make important contributions to our planet. Many animals depend on flies for food. Housefly pupae contain over 50 percent protein and high amounts of fat. Scientists estimate that housefly pupae may be even more nutritious than soybeans. And flies multiply so quickly that scientists are now looking for ways to grow, harvest, and dehydrate flies for animal feed.

In addition, flies are essential to keeping our world clean. If these insects didn't exist, we would be surrounded by decaying animals and plants. Flies, especially in their larval stages, eat at enormous rates, breaking down manure and dead matter into nearly odorless materials that enrich the soil.

Scientists are investigating ways to put these trash eaters to use in disposing of garbage and human sewage. In the future, we may find that the small housefly is a partial answer to our world's big waste problem.

As you have seen, the housefly may not be as beautiful as the butterfly, or as productive as the honeybee, but it is an important and interesting part of our world. So the next time you spy one of these flying gymnasts, think about how much it contributes to nature's careful balance—and then shoo it away from your food!

GLOSSARY

abdomen: the third and rear portion of an insect's body

aedeagus: part of the male housefly's external reproductive organ

antennae (singular *antenna*): the paired sense organs extending from the heads of insects

compound eyes: eyes made up of many single eyes that have their own nerves, pigment, and lenses

chitin: one of the substances that makes an insect's exoskeleton hard

crop: an enlarged sac that is connected to an insect's digestive tract and stores food

diapause: a temporary stopping of a housefly's body functions that is similar to hibernation

egg: a female reproductive cell

exoskeleton: the hard, protective covering over an insect's body

femur: the third and middle segment of a housefly's leg

fertilization: the joining of sperm and egg

halteres (singular *haltere*): a pair of small club-shaped organs, located behind a true fly's wings, that are used for flying balance and control

labella (singular *labellum*): the spongy pads at the end of the housefly's proboscis

larvae (singular *larva*): the immature insects that hatch from eggs. They are commonly called maggots.

metamorphosis: the transformation of an insect from an egg to an adult

molt: to shed an outer layer of skin

ocelli (singular *ocellus*): small simple eyes found on some insects' heads

ovipositor: a female housefly's external reproductive organ

palpi (singular *palpus*): the paired sensory organs extending from the top of the housefly's proboscis

proboscis (plural *proboscises*): the long sucking tube that the housefly extends to eat

pulvilli (singular *pulvillus*): the hairy pads at the end of a housefly's tarsi

pupae (singular *pupa*): the immature insects, enclosed in cases or cocoons, that have developed from larvae and are changing into adults

species: a scientific grouping of living things that share similar characteristics

sperm: male reproductive cells

spiracles: the external openings in an insect's respiratory system

tarsi (singular *tarsus*): the fifth and lowest segment of a housefly's leg. It is composed of five parts.

thorax: the second and middle section of an insect's body

tibia: the fourth and second lowest segment of the housefly's leg

INDEX

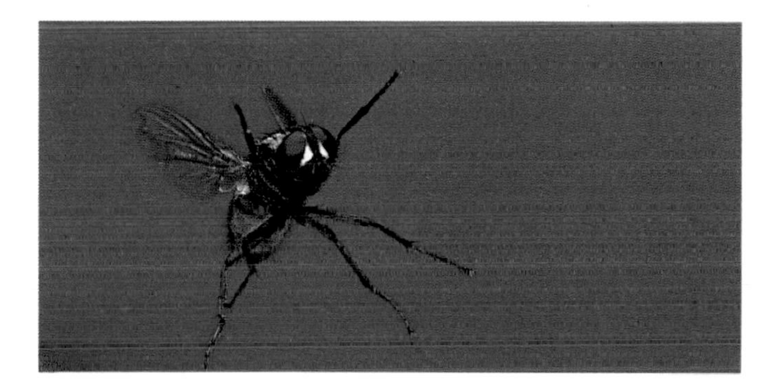

ABOUT THE AUTHORS

Heiderose and Andreas Fischer-Nagel received degrees in biology from the University of Berlin. Their special interests include animal behavior, wildlife protection, and environmental control. The Fischer-Nagels have collaborated as authors and photographers on several internationally successful science books for children. They attribute the success of their books to their "love of children and of our threatened environment" and believe that "children learning to respect nature today are tomorrow's protectors of nature."

The Fischer-Nagels live in Germany with their daughters, Tamarica and Cosmea Désirée.

Additional photographs courtesy of: Gay Bumgarner, PHOTO/NATS, p. 2; Nelson-Bohart & Associates, p. 3, 20; James H. Robinson, p. 17, 26; Edward S. Ross, right column, p. 43. Switzerland's Natural History Photography Agency, p. 5, right column, 22. Front cover courtesy of James H. Robinson. Diagrams p. 14, 15 by Laura Westlund.